はしがき

　このペットボトルの蓋は、蓋を成形する時に同時に指かかりを設ける。使用に当たっては、左手でペットボトルを支え、右手でペットボトルの蓋を握り、左右に回すことにより、ペットボトルの蓋を容易に開閉できる。

　汗等で滑ることなく、また、力の弱い子供、高齢者、女性でも容易にペットボトルの蓋が開けることが出来る。これにより、片手で容易に使用している著作物を提供する。

Preface

　The lid of this PET bottle can open and close the lid of a PET bottle easily by forming a convex type in the side, supporting a PET bottle with the left hand, grasping the lid of a PET bottle with the right hand in use, and turning to right and left.

　The lid of a PET bottle can also open the weak child of power, elderly people, and a woman easily, without sliding with sweat etc.

　Illustration copyright of the directions for use which turn the lid of a PET bottle single hand

目　次

1、微力で回せるペットボトルキャップの使用方法（イラスト解説）----------6

(1) 回らないで困った表情と簡単に回ったのでニコニコした表情------------------6

(2) 幼稚園児の男の子キャップを回せる表情----------------------------------7

(3) 小学生の女の子が回せる表情--8

(4) 小学生の男の子が回せる表情--9

(5) 手が不自由なおばさん、背景の介護犬が喜んでいる表情　開いたらﾜﾝ・ﾜﾝ--10

(6) 病院のベットの上で、手が不自由なおじいさん、片手でｷｬｯﾌﾟを回せる表情--11

(7) 片手を怪我で吊り下げて包帯をしている少年が膝と膝の間に挟んで回す表情-12

(8) 力が無い子供の男の子ｷｬｯﾌﾟを回せる　背景のﾈｺのうれしそうな表情--------13

(9) 釣堀で片手で釣り竿を握りながら、片手でキャップを回せる表情-----------14

(10) 自動車を運転中に片手でキャップを回せる表情--------------------------15

(11) 膝に挟んでキャップを回せる表情-------------------------------------16

(12) 読書をしながら片手でキャップを回せる表情----------------------------17

(13) 脇の下に挟んでキャップを回せる表情---------------------------------18

(14) 固定電話中に片手でキャップを回せる表情------------------------------19

(15) 携帯電話中に片手でキャップを回せる表情------------------------------20

(16) パチンコをしながら片手でキャップを回せる表情-------------------------21

English of the usage

2、

The directions for the PET bottle cap which can be turned in poor ability

(1)--23

Since it turned simply, I am glad.   It was troubled without turning.

(2)--24

A kindergartener's boy can also turn a cap.

(3)--25

A schoolchild's girl can also turn a cap.

(4)--26

A schoolchild's boy can turn a cap.

(5)--27

A hand can turn an inconvenient lady cap.

(6)--28

On the bed of a hospital, a hand can turn the cap of an inconvenient grandfather and a PET bottle single hand.

(7)--29

The boy who hangs one hand by injury and is doing the bandage inserts between knees, and can turn a cap.

(8)--30

A weak child's boy cap can be turned.
The cat of a background seems to be glad.

(9)--31

By fishing, I can turn the cap of the plastic bottle with one hand.

(10)--32
I am driving a car and can turn the cap of the plastic bottle with one hand.

(11)--33
Between knees, a PET bottle is inserted and a cap can be turned.

(12)--34
I can turn the cap of the plastic bottle with one hand while reading.

(13)--35
I sandwich the plastic bottle between an arm and chests and can turn a cap.

(14)--36
I can turn the cap of the plastic bottle with one hand in a landline.

(15)--37
In cell-phones, I can turn the cap of the plastic bottle with one hand.

(16)--38
I can turn the cap of the plastic bottle with one hand while playing pachinko.

3、公報解説--39

4、Patent journal English--44

1、微力で回せるペットボトルキャップの使用方法（イラスト解説）

(1) 回らないで困った表情と簡単に回ったのでニコニコした表情

(2) 幼稚園児の男の子キャップを回せる表情

(3) 小学生の女の子が回せる表情

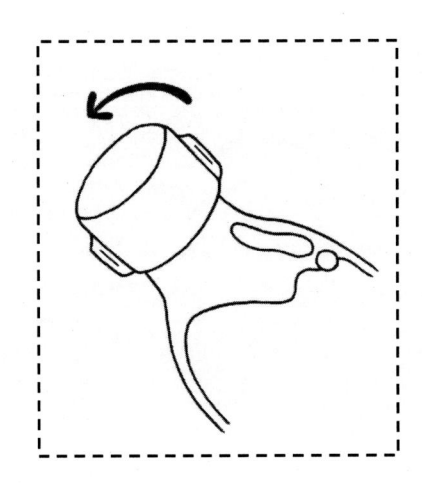

(4) 小学生の男の子が回せる表情

(5) 手が不自由なおばさん、背景の介護犬が喜んでいる表情　開いたらワンワン

⑹　病院のベットの上で、手が不自由なおじいさん　片手でキャップを回せる表情

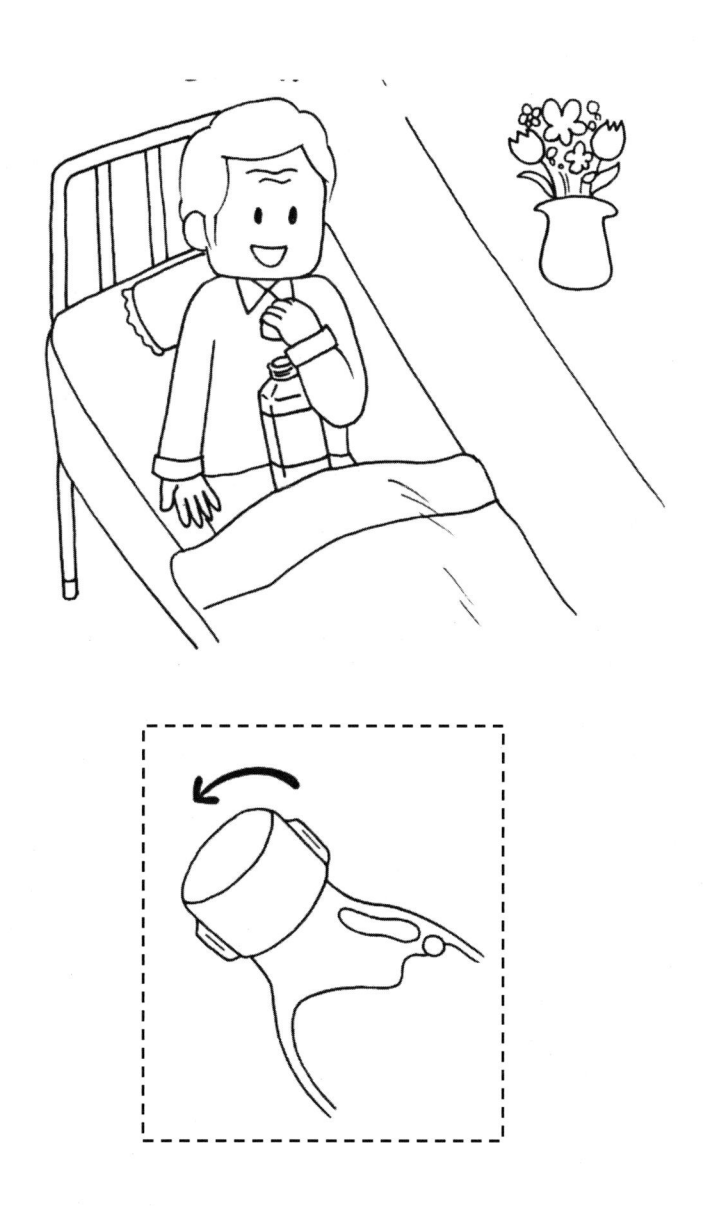

⑺　片手を怪我で吊り下げて包帯をしている少年が膝と膝の間に挟んで回す表情

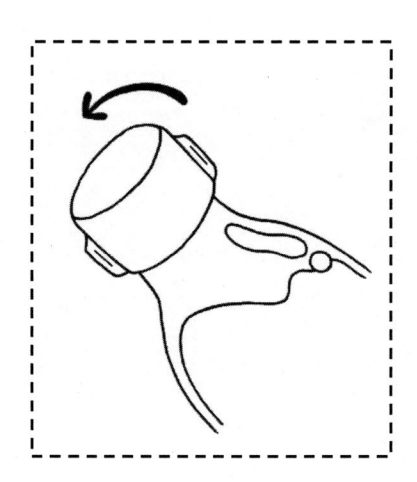

⑻　力が無い子供の男の子キャップを回せる　背景のネコのうれしそうな表情

⑼　釣堀で片手で釣り竿を握りながら、片手でキャップを回せる表情

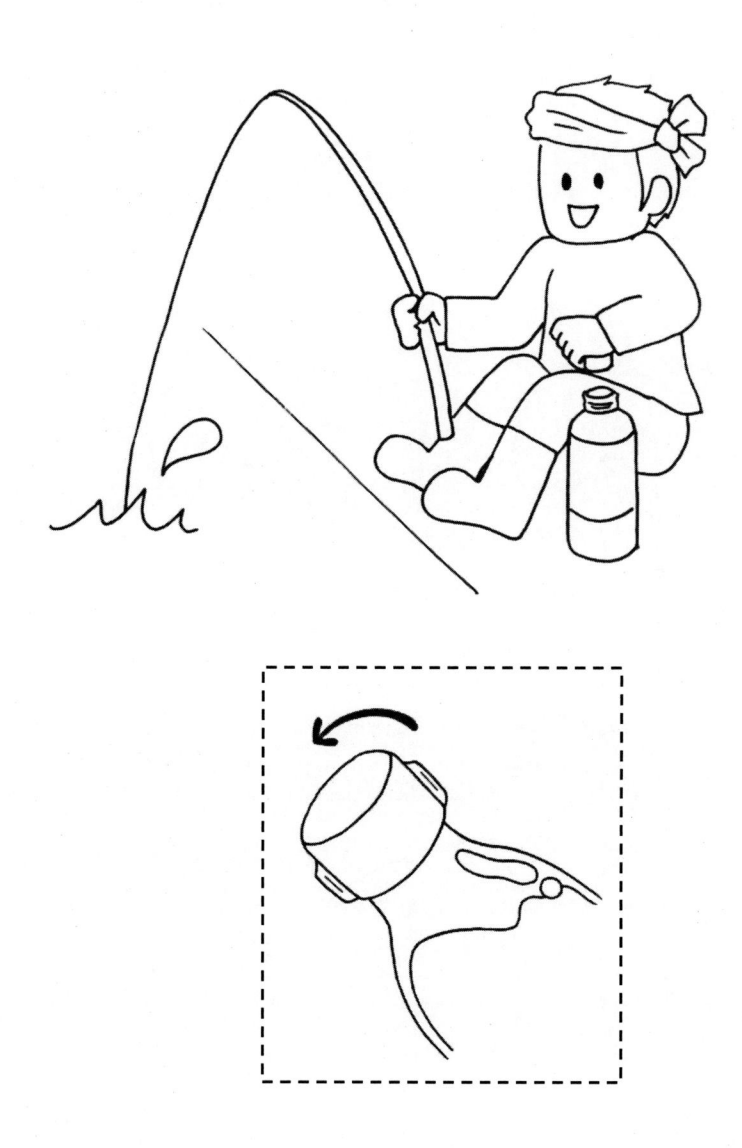

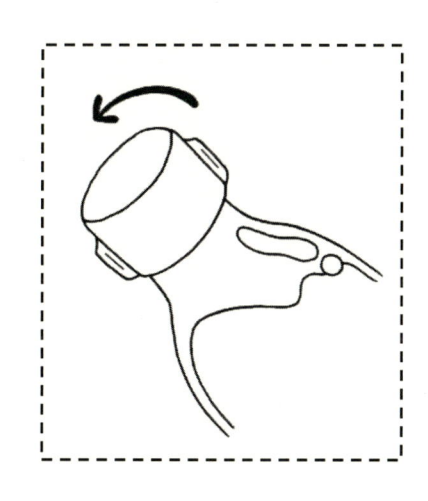

(10)　自動車を運転中に片手でキャップを回せる表情

(11)　膝に挟んでキャップを回せる表情

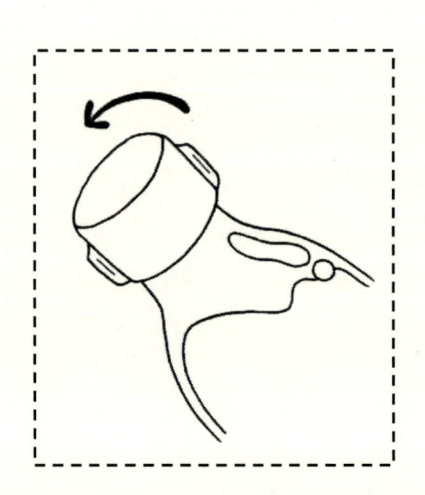

⑿　読書をしながら片手でキャップを回せる表情

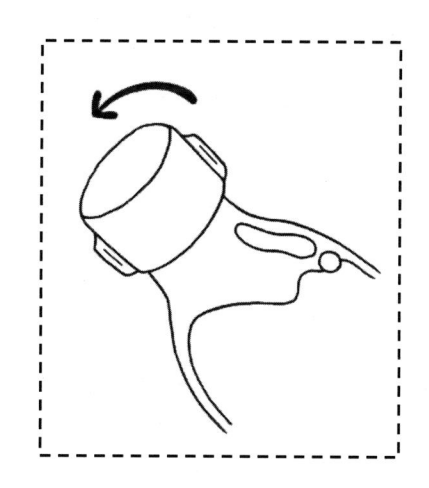

⑬　脇の下に挟んでキャップを回せる表情

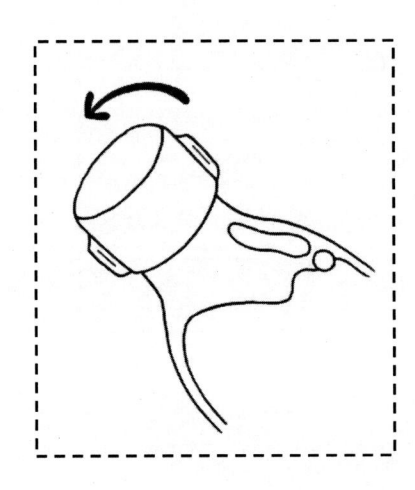

⑭　固定電話中に片手でキャップを回せる表情

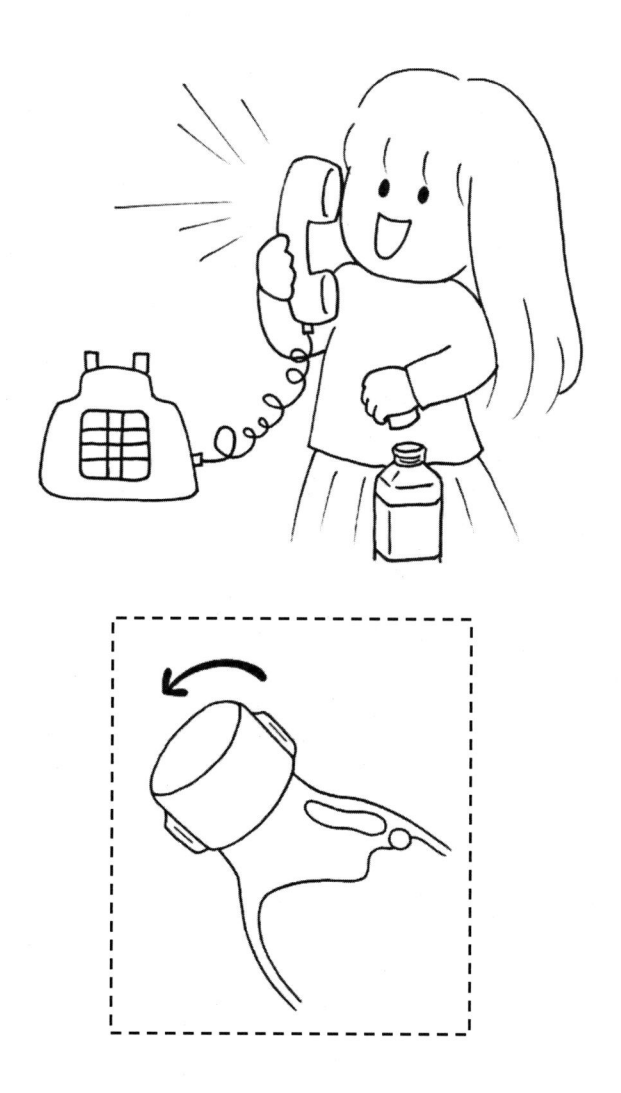

⒂　携帯電話中に片手でキャップを回せる表情

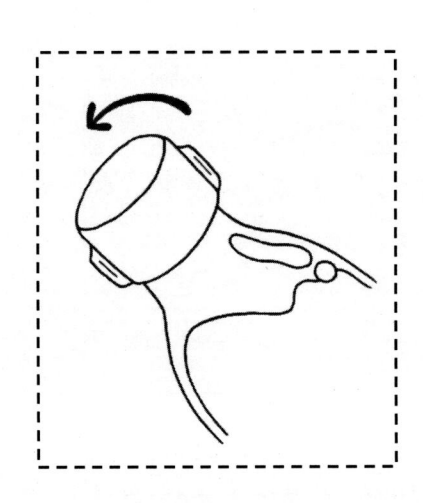

⒃　パチンコをしながら片手でキャップを回せる表情

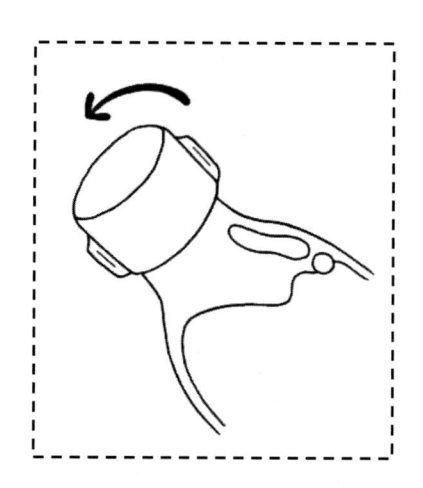

English of the usage

2

The directions for the PET bottle cap which can be turned in poor ability
(1)

Since it turned simply, I am glad.　　　　It was troubled without turning.

(2)
A kindergartener's boy can also turn a cap.

(3)
A schoolchild's girl can also turn a cap.

(4)
A schoolchild's boy can turn a cap.

(5)
A hand can turn an inconvenient lady cap.

(6)
On the bed of a hospital, a hand can turn the cap of an inconvenient grandfather and a PET bottle single hand.

(7)
The boy who hangs one hand by injury and is doing the bandage inserts between knees, and can turn a cap.

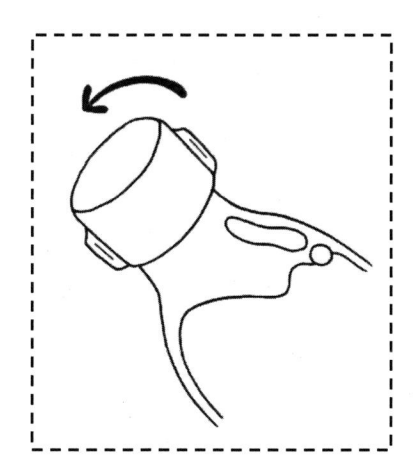

(8)
A weak child's boy cap can be turned.
The cat of a background seems to be glad.

(9)
By fishing, I can turn the cap of the plastic bottle with one hand.

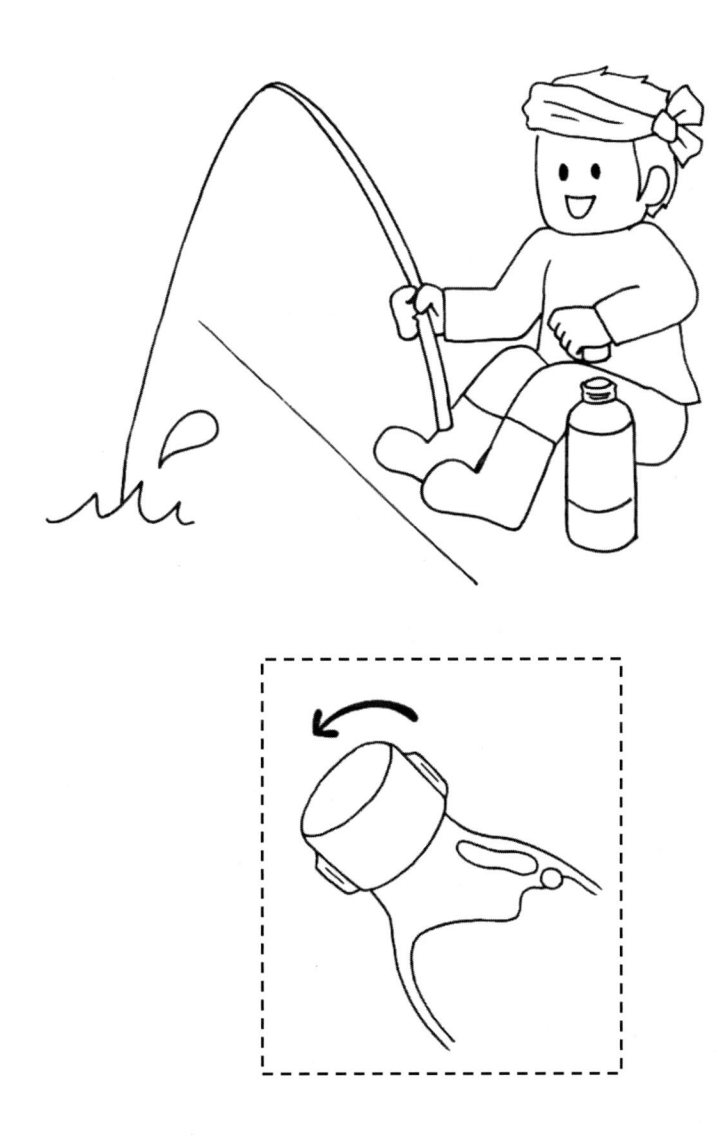

(10)
I am driving a car and can turn the cap of the plastic bottle with one hand.

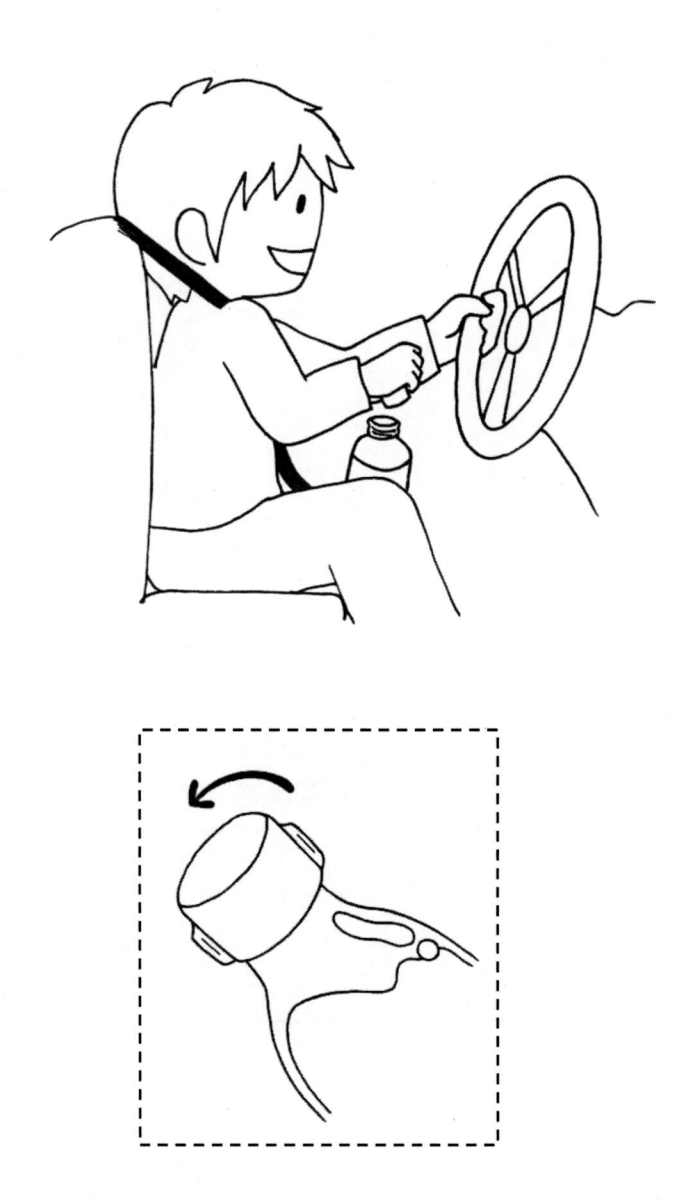

(11)
Between knees, a PET bottle is inserted and a cap can be turned.

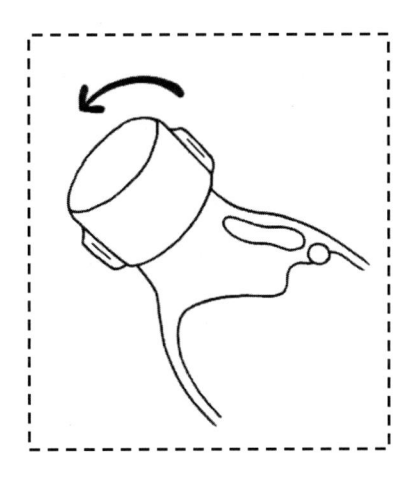

(12)
I can turn the cap of the plastic bottle with one hand while reading.

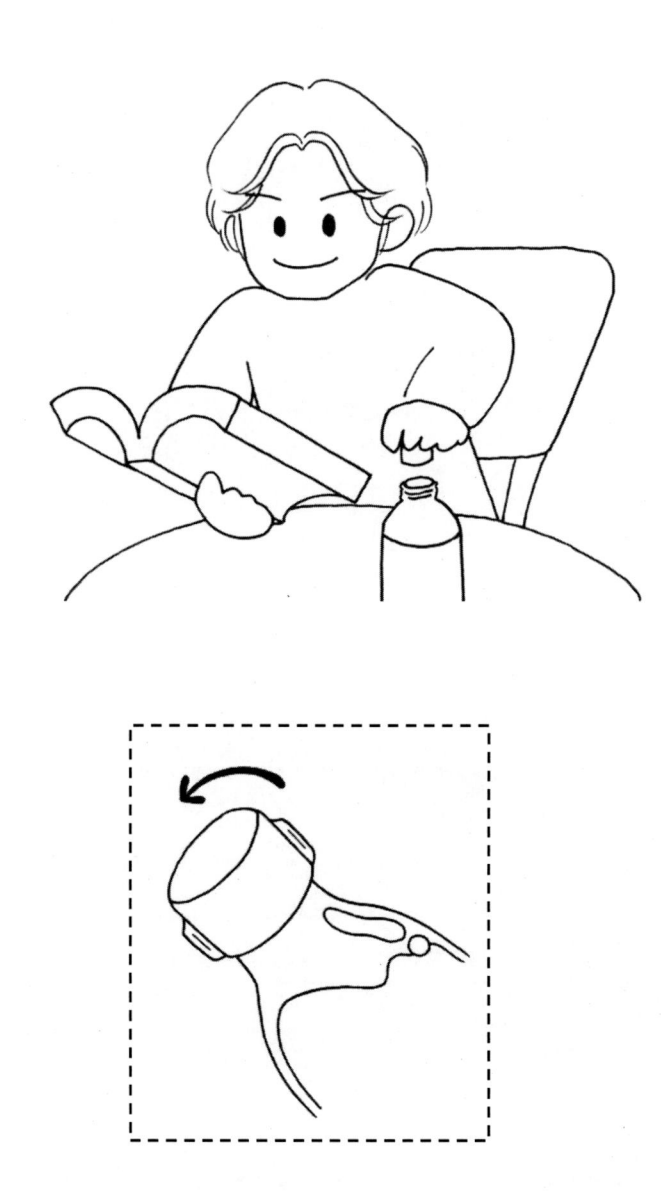

(13)
I sandwich the plastic bottle between an arm and chests and can turn a cap.

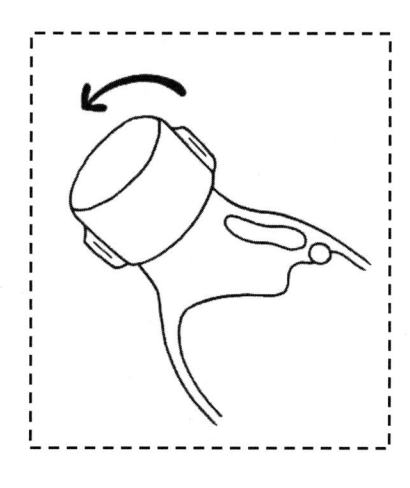

(14)
I can turn the cap of the plastic bottle with one hand in a landline.

(15)
In cell-phones, I can turn the cap of the plastic bottle with one hand.

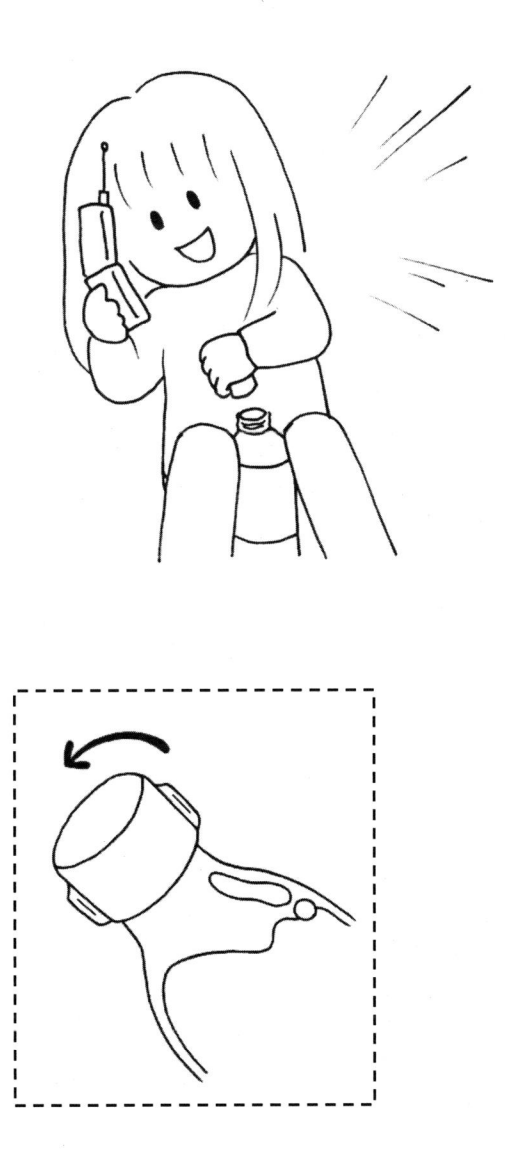

(16)
I can turn the cap of the plastic bottle with one hand while playing pachinko.

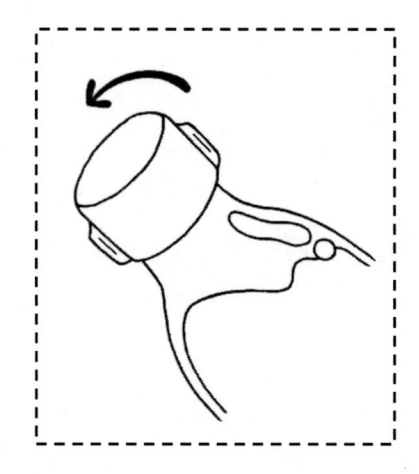

3、公報解説

(1) 実用新案登録第３１７６６２２号

　　考案の名称；ペットボトルの蓋

　　実用新案権者；岡村タカ子

　　考案者；藤井勝允

【要約】
【課題】汗等で滑ることなく、また、力の弱い子供、高齢者、女性でも容易にペットボトルの蓋が開けることが出来るペットボトルの蓋を提供する。
【解決手段】ペットボトルＣの蓋Ａに蓋を成形する時に同時に指かかりＢを設ける。使用に当たっては、左手でペットボトルＣを支え、右手でペットボトルＣの蓋Ａを握り、左右に回すことにより、ペットボトルＣの蓋Ａを容易に開閉できる。
【選択図】図１
【実用新案登録請求の範囲】
【請求項１】
飲料水その他のペットボトルの蓋に、指かかりを設けた蓋。
【考案の詳細な説明】
【技術分野】
【０００１】
本考案は、ペットボトルの蓋部分に指かかりを設けた蓋である。
【背景技術】
【０００２】
従来のペットボトルの蓋は小さく、汗等で滑る為、力の弱い子供、高齢者、女性な

どにはペットボトルの蓋が開けにくかった。

【考案の概要】

【考案が解決しようとする課題】

【０００３】

今までのペットボトルの蓋は小さく、汗等で滑る為、力の弱い子供、高齢者、女性にはペットボトルの蓋が開けにくかった。

【課題を解決するための手段】

【０００４】

ペットボトルの蓋部分に、指かかりを設ける。

【考案の効果】

【０００５】

ペットボトルの蓋部分に、指かかりを設けることにより、ペットボトルの蓋を開ける時に指が指かかりにかかり、滑らずに弱い力でも容易にペットボトルの蓋を開けることが可能となる。

【図面の簡単な説明】

【０００６】

【図１】Ｃはペットボトル本体、Ｄはペットボトル口部、Ａはペットボトルの蓋、Ｂはペットボトルの蓋部分に指かかりを設けた部分の実施例を示す一部切欠断面図の立面図である。

【図２】Ｃはペットボトル本体、Ｄはペットボトル口部、Ａはペットボトルの蓋、Ｂはペットボトルの蓋部分に指かかりを設けた部分の実施例を示す一部切欠断面図の平面図である。

【考案を実施するための形態】

【０００７】

ペットボトルの蓋に指かかりを設け、仮に左手でペットボトルを支え、右手でペットボトルの蓋を握り、左右に回すと、ペットボトルの蓋が容易に開閉できる。

【実施例】

【０００８】

以下添付図面に従って、実施例を説明する。

Cは、ペットボトル本体である、Dはペットボトル口部、Aはペットボトルの蓋、Bはペットボトルの蓋部分に指かかりを設けたペットボトルの蓋である。

【産業上の利用可能性】

【０００９】

これまで、非力などで利用を控えていた隠れた消費者に対して、需要の掘り起こしが見込まれ、消費量の増大を見込める事と、ペットボトルの蓋に指かかりを付加する事は簡単な成型技術で対応できるため利用価値は高いものと想定する。

また、これまで開けられなかった消費者の尊厳の回復にも寄与するものと思慮する。

【符号の説明】

【００１０】

A　ペットボトルの蓋

B　ペットボトルの蓋に指かかりを設けた部分

C　ペットボトル本体

D　ペットボトルの蓋の口部

【図面の簡単な説明】

【０００６】

【図１】Cはペットボトル本体、Dはペットボトル口部、Aはペットボトルの蓋、

Bはペットボトルの蓋部分に指かかりを設けた部分の実施例を示す一部切欠断面図の立面図である。

【図2】Cはペットボトル本体、Dはペットボトル口部、Aはペットボトルの蓋、Bはペットボトルの蓋部分に指かかりを設けた部分の実施例を示す一部切欠断面図の平面図である。

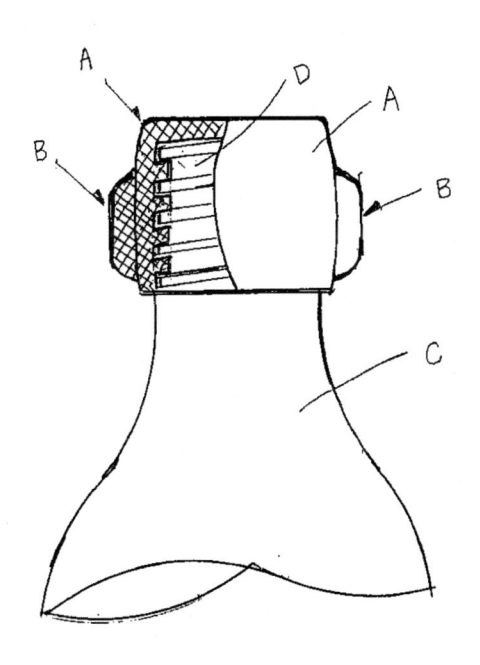

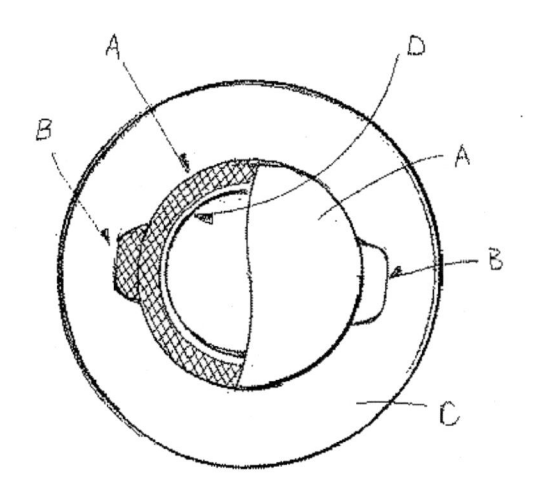

4. Patent journal English

DETAILED DESCRIPTION

[Detailed explanation of the device]

[Field of the Invention]

[0001]

This design is the lid which provided finger starting to the lid part of the PET bottle.

[Background of the Invention]

[0002]

The lid of the conventional PET bottle was small, and since it slid with sweat etc., the lid of the PET bottle could not open it in a child with weak power, elderly people, and a woman easily.

[The outline of a device]

[Problem(s) to be Solved by the Device]

[0003]

The lid of the old PET bottle was small, and since it slid with sweat etc., the lid of the PET bottle could not open it in a child with weak power, elderly people, and a woman easily.

[Means for solving problem]

[0004]

To the lid part of a PET bottle, it provides finger starting.

[Effect of the Device]

[0005]

By providing finger starting to the lid part of a PET bottle, when opening the lid of a PET bottle, a finger becomes possible [starting finger starting and opening the lid of a PET bottle easily also by weak power, without sliding].

[Brief Description of the Drawings]

[0006]

[Drawing 1]C shows the working example of the portion to which the main part of a PET bottle and D provided the PET bottle regio oralis, A provided the lid of the PET bottle, and B provided finger starting to the lid part of the PET bottle -- it is an elevational view of a notching cross sectional view in part.

[Drawing 2]C shows the working example of the portion to which the main part of a PET bottle and D provided the PET bottle regio oralis, A provided the lid of the PET bottle, and B provided finger starting to the lid part of the PET bottle -- it is a plan view of a notching cross sectional view in part.

[The form for devising]

[0007]

If provide finger starting on the lid of a PET bottle, a PET bottle is supported with the left hand temporarily, the lid of a PET bottle is grasped with the right hand and it turns to right and left, the lid of a PET bottle can open and close easily.

[Working example]

[0008]

According to an accompanying drawing, an working example is described below. It is a lid of the PET bottle in which D provided the PET bottle regio oralis, A provided the lid of the PET bottle, and B provided finger starting to the lid part of the PET bottle whose C is a main part of a PET bottle.

[Industrial applicability]

[0009]

Since it can respond with easy molding technology, utility value assumes it to be high things that digging up of demand is expected and increase of an amount of consumption can be expected until now to the hidden consumer who was refraining from use powerlessly etc., and to add finger starting to the lid of a PET bottle.

It considers with contributing also to dignified recovery of the consumers who were not able to open until now.

[Explanations of letters or numerals]

[0010]

A The lid of a PET bottle

The portion which provided finger starting on the lid of B PET bottle

C The main part of a PET bottle

Regio oralis of the lid of D PET bottle

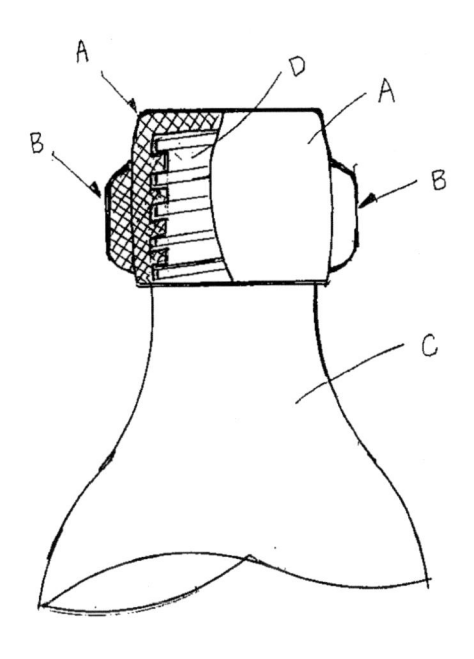

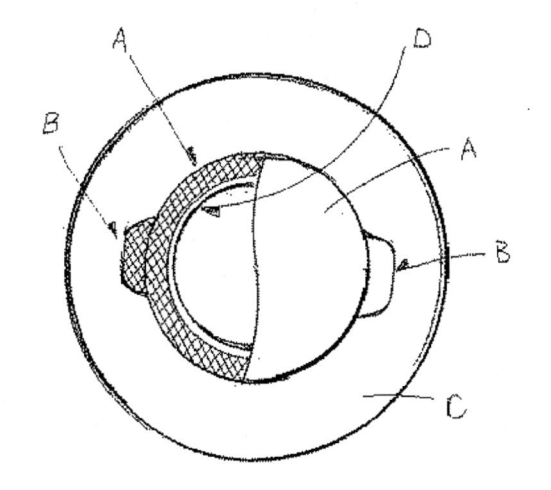

力の弱い子供でも回せる・片手でも回せるペットボトルのキャップ

定価（本体1,000円＋税）

２０１３年（平成２５年）６月２８日発行

No. OK-016

発行所　発明開発連合会®

東京都渋谷区渋谷2-2-13

電話 03-3498-0751㈹

発行人　ましば寿一

著作権企画　発明開発連合会

Printed in Japan

著者　岡村タカ子 ©

本書の一部または全部を無断で複写、複製、転載、データーファイル化することを禁じています。

It forbids a copy, a duplicate, reproduction, and forming a data file for some or all of this book without notice.